Cows

written by Mia Coulton
photographed by
Mia Coulton & Hannah Brown

This is a cow.

3

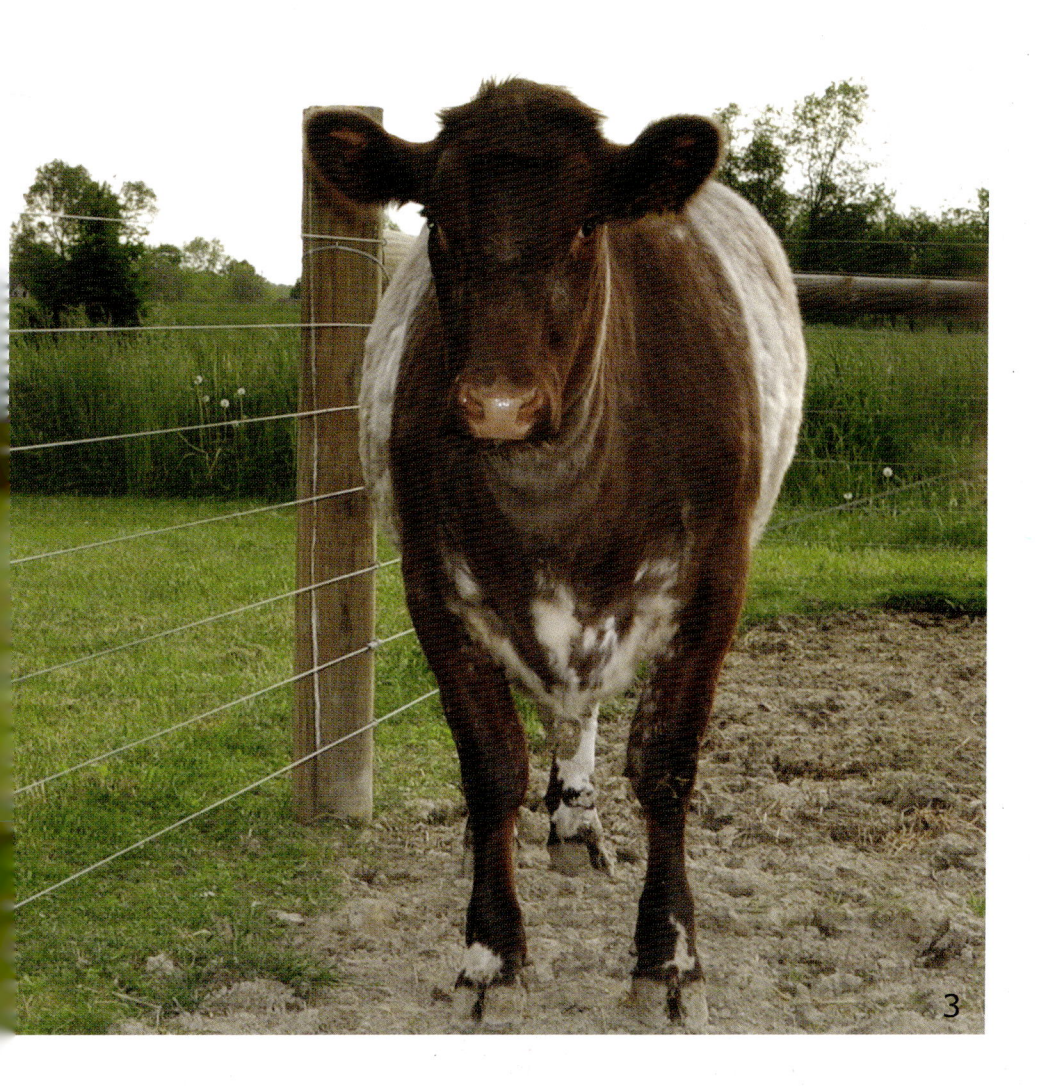

This cow is eating hay.

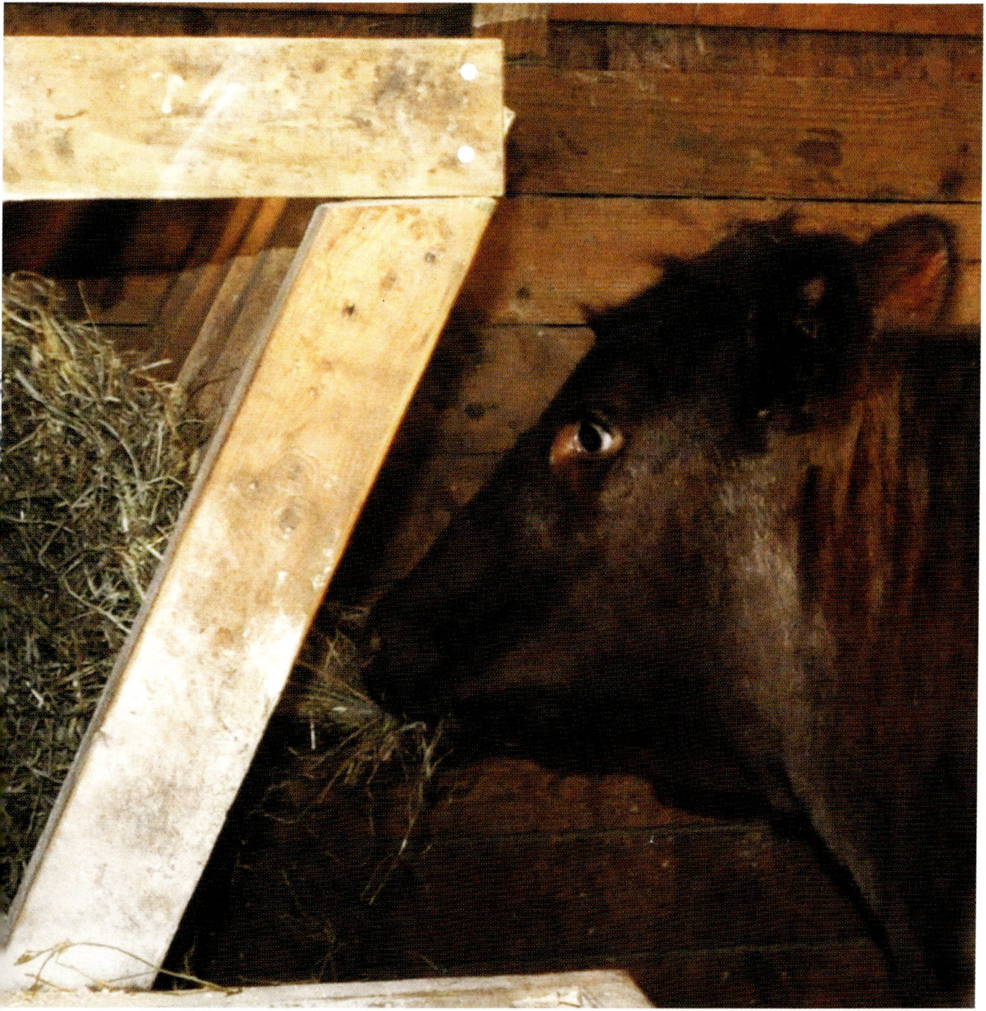

This cow is eating grass.

7

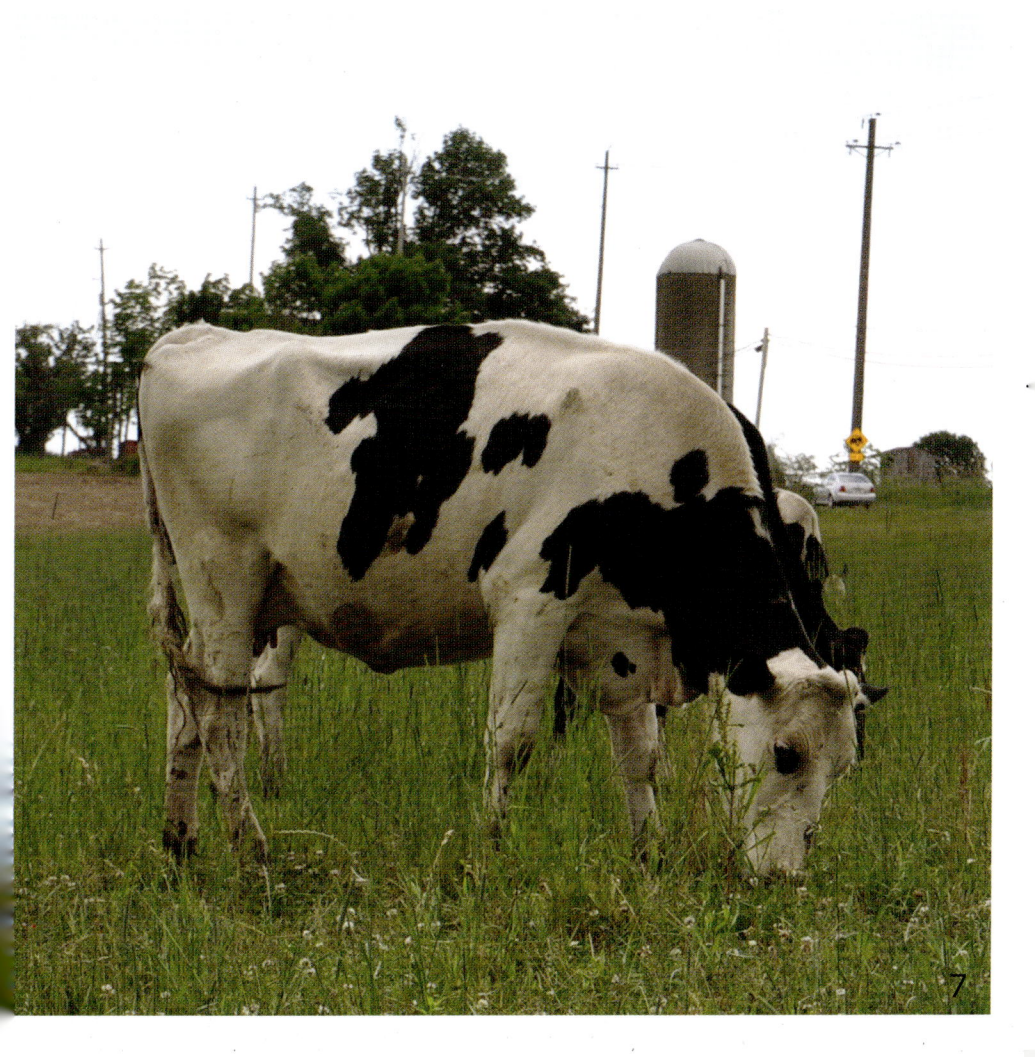

This cow is running.

A cow can run fast.

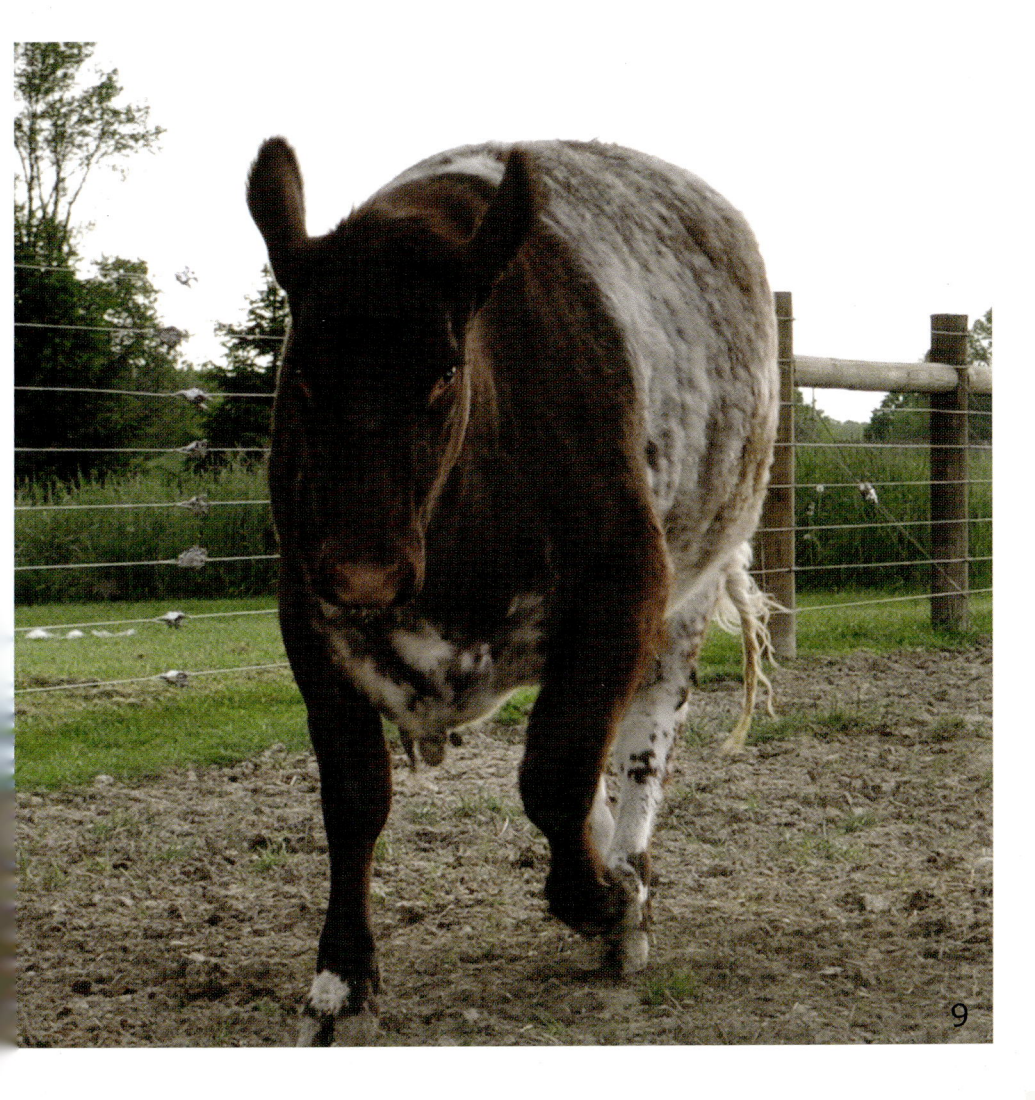

This cow and her baby are resting.

A baby cow is a calf.

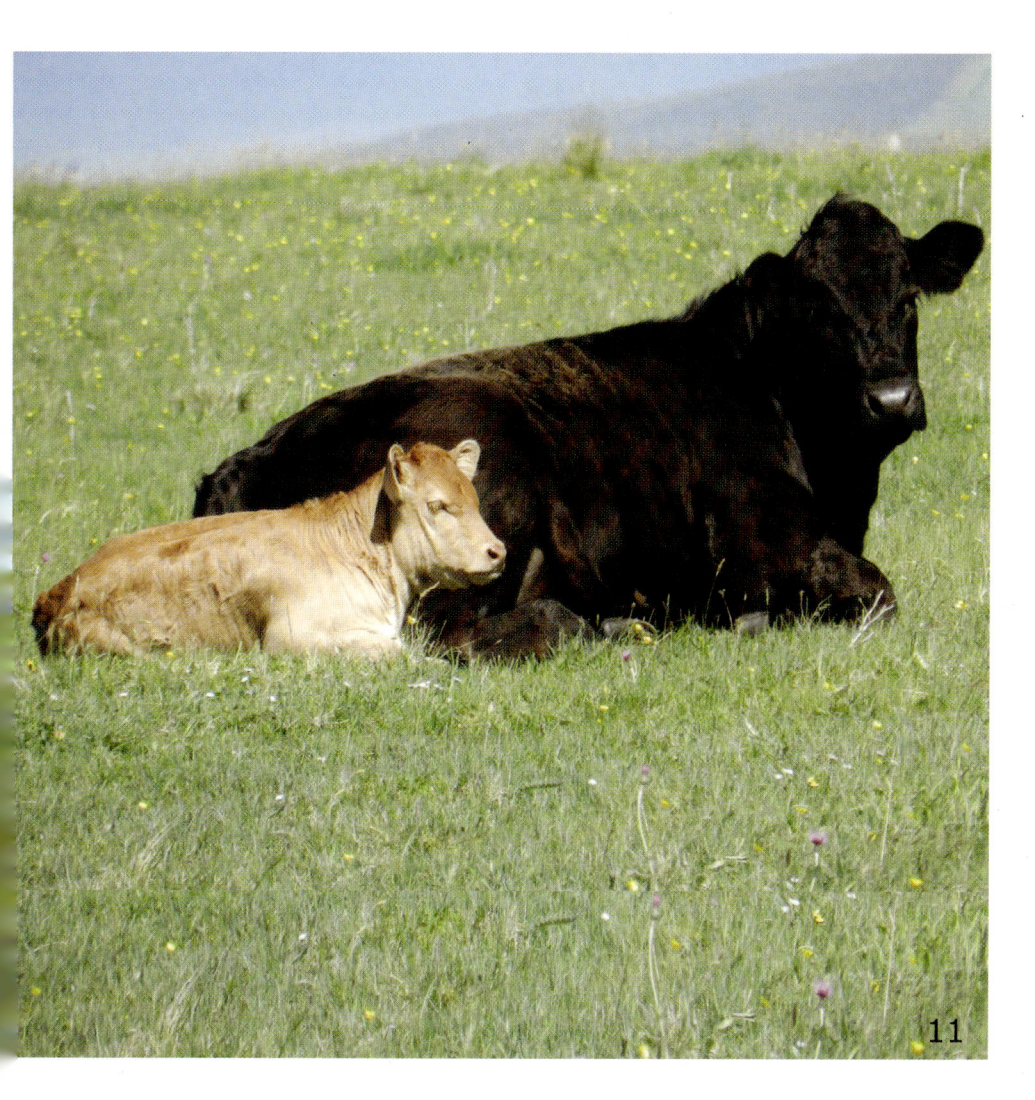

"Moo."

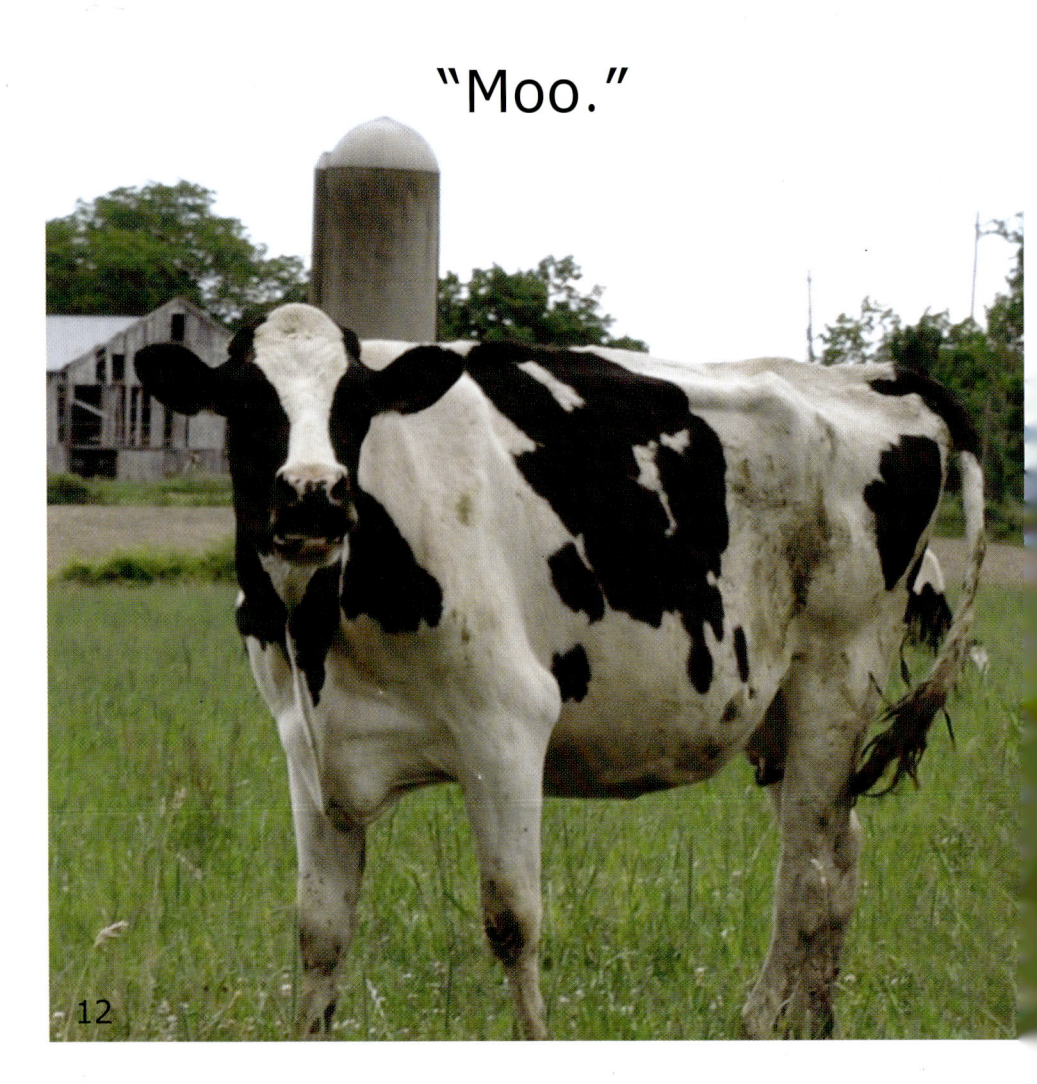